WHAT ABOUT SCIENCE

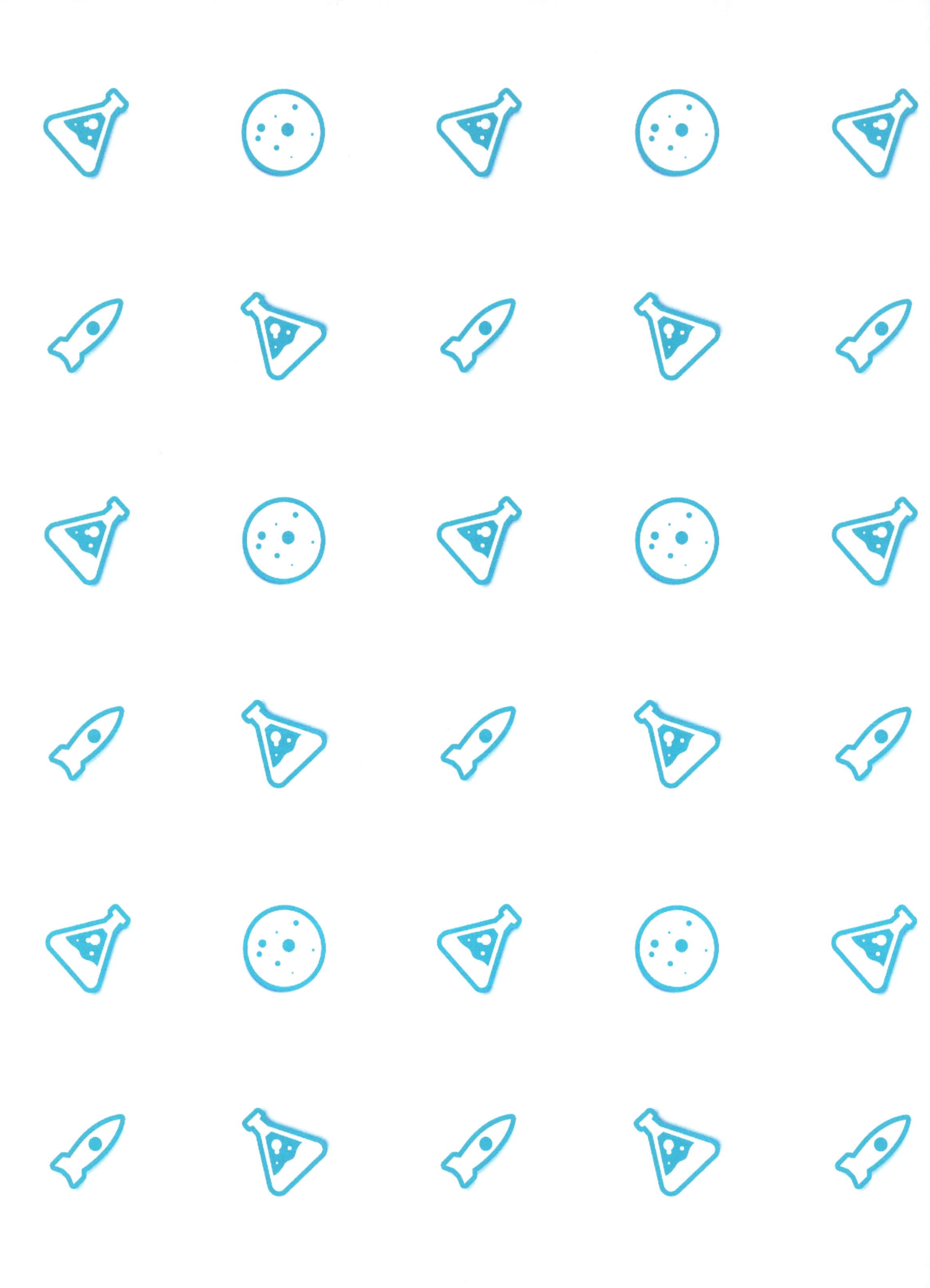

WHAT ABOUT SCIENCE

by Eugene Fedorenko

Copyright © 2018 Eugene Fedorenko
All rights reserved.

ISBN: 9781793024220

DEDICATION

Dedicated to all true scientists who work for the humanity' prosperity and future.

Trust me - such people are still among us.

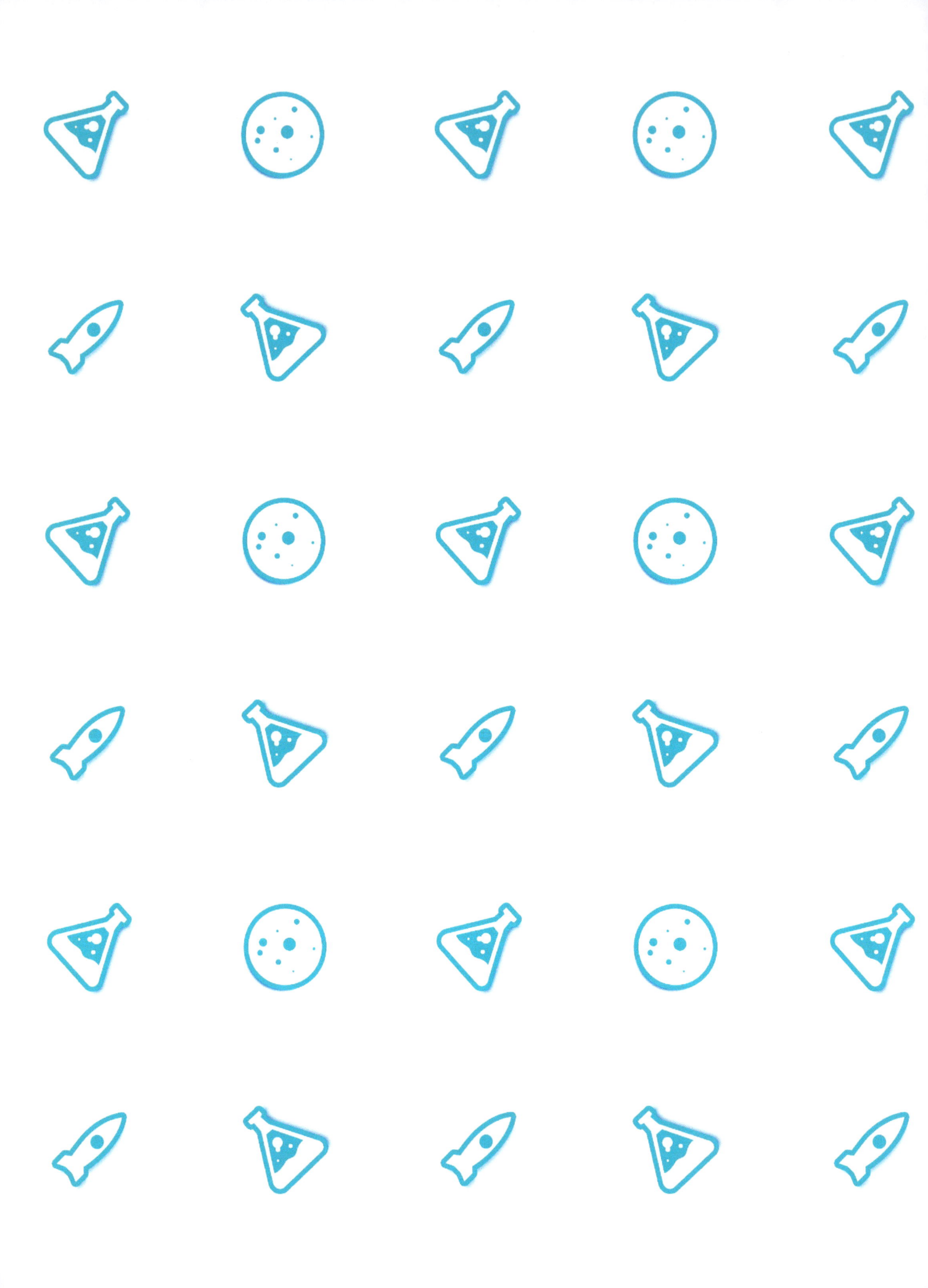

TABLE OF CONTENTS

1. CO2 - the greenhouse gas - can be removed from the athmosphere and... turned into rock

2. Researchers created a list with top causes of stroke worldwide, air pollution included

3. How to play games without any physical controllers? Right - using your brain only

4. New hope rises for patients with severe multiple sclerosis (MS) diagnosis

5. What can we learn from the large Antarctic iceberg that had recently calved away?

6. What do you think about scientific facts? It appears your opinion actually matters

7. NHTSA's NMVCC survey - 2-3 hours of sleep lead to doubled risk for a crash

8. New Zealand's big geological problem - now explained

9. New optical wireless internet connection could replace Wi-Fi in 5 years

10. How do fears develop in children? It appears they can be adopted from parents and friends

13. Will you help others in life-threatening situations first? Or maybe yourself?

14   Hit your 60th? No problem - you can still become a programmer

15   Are you an impulsive person? If so, you might want to stay away from drugs

16   Even a few nuclear strikes will cause severe climate changes around the world

17   Did you see those 4k or even 10k displays? Multiply these resolutions by a factor of 3

18   How can Vitamin D be possibly associated with autism and pregnant women?

19   Can we predict major tsunamis' occurrence, or can't we?

20   Bots can affect news on the social networks - with both positive and negative impact

21   How does air temperature affect sleep quality?

22   The Atlantic Meridional Overturning Circulation (AMOC) is at risk of disruption

23   Our brain unconsciously misjudges speed after it gets used to the current one

24   Thunderstorms occur twice as frequenty over the areas with high aerosol emissions

25   New AI that simulates social interaction in computer games is developed

CO_2 - the greenhouse gas - can be removed from the athmosphere and... turned into rock

QUICK BENEFITS: Pumping something into volcanic layers may affect seismic activity - try to avoid visiting such areas

An international group of scientists developed a promising method of carbon dioxide (CO_2) disposal.

The main principle is to dissolve CO_2 in water and carry it down into the volcanic rock layers through the pre-drilled well.

When the fluid reaches the rock layers - basalt minerals to be clear -, CO_2 starts to react actively, forming carbonate minerals, essentially environment-friendly rocks.

© https://www.carbfix.com

© https://www.carbfix.com

The research is called CarbFix and is funded by icelandic, european and US institutes and environmental organisations.

Original source: University of Southampton
Year: 2016

Researchers created a list with top causes of stroke worldwide, air pollution included

QUICK BENEFITS: Lead a healthy lifestyle > At home, minimize chemicals usage & use air purifiers > Live in eco-friendly places

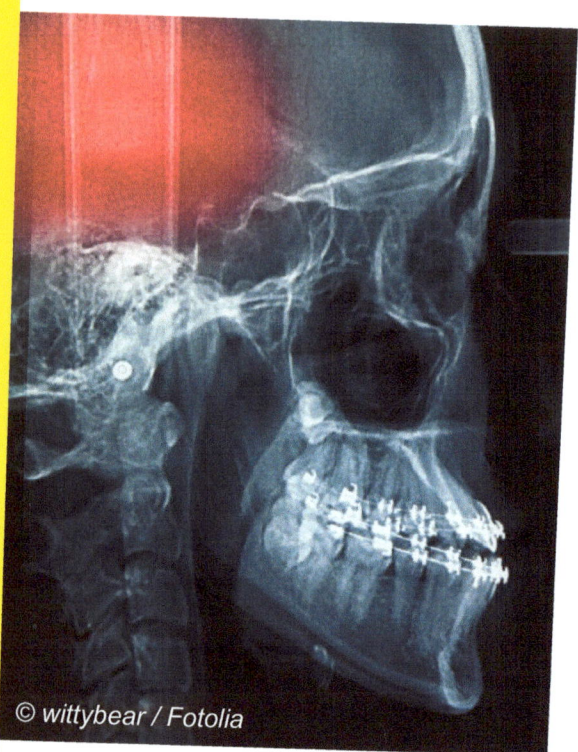

After the analysis of the data from the Global Burden of Disease Study held from 1990 to 2013, researchers could create a list of top known causes of stroke in 188 countries.

The 10 leading factors for stroke were: high blood pressure, diet low in fruit, high body mass index (BMI), diet high in sodium, smoking, diet low in vegetables, environmental air pollution, household pollution from solid fuels, diet low in whole grains and high blood sugar.

The priority of mentioned factors varies between developing and developed countries.

For the first time, air pollution was listed as a risk factor for stroke development.

Original source: The Lancet Neurology
Year: 2016

How to play games without any physical controllers? Right - using your brain only

QUICK BENEFITS: Expect smart prosthetics, new VR appliances and, who knows, mind readers? Stay sharp

Scientists developed a method for non-invasive interaction with virtual reality via direct brain stimulation.

They generated a visual stimulation artifact - phosphene - directly in the brain using transcranial magnetic stimulation.

If put simply - you can now play computer games without using any physical controls.

© http://neubay.com

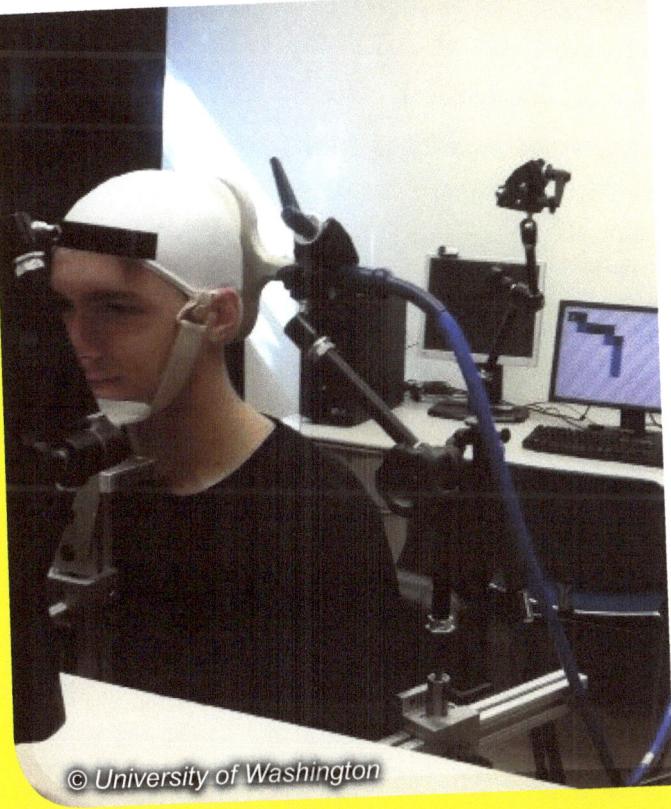
© University of Washington

Current accuracy of brain signal tracing using mentioned technic reaches 92%.
The technology is still in development - scientists strive to make it portable

Original source: University of Washington
Year: 2016

New hope rises for patients with severe multiple sclerosis (MS) diagnosis

QUICK BENEFITS: If you or your relative got an MS - search for stem-cell-powered immune system replacement therapies

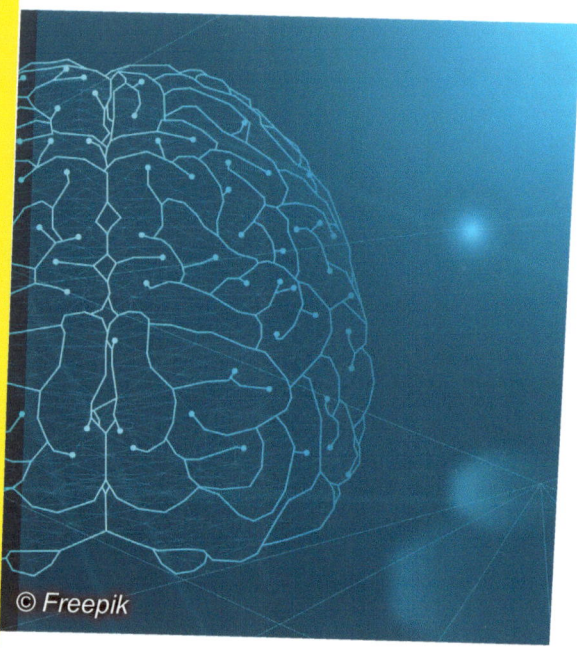

New hope rises for patients with severe multiple sclerosis diagnosis

It is known that multiple sclerosis (MS) usually leads to early brain functions decline and death.

Scientists developed a perspective approach to treating MS, which actually is very similar to the one used for decades for treating some forms of leukemia.

The procedure looks like that:

- patient bone's stem cells are extracted and stored while his immune system is suppressed with strong drugs;
- stem cells are injected back into the patient restoring the immune system without its "memory" of attacking his own body.

The new approach is quite effective and, in most cases, leads to lasting decline in the disease progression or even to its complete stop.

However, this procedure has significant side effects by itself, up to lethal outcome. Because of that, it should be administered only to those with early and severe MS cases.

Original source: Ottawa Hospital Research Institute
Year: 2016

What can we learn from the large Antarctic iceberg that had recently calved away?

QUICK BENEFITS: If looking into the future, you might want to relocate to higher and, preferably, secure south areas

More than a year ago one of the largest recorded icebergs A-68 had calved from the Larsen C Antarctic ice shelf.

It was originally twice the size of Luxembourg and 12% of the whole Larsen C ice shelf size.

Such declines in Antarctic ice cover may eventually lead to global sea level rise and changes in ocean currents' directions due to the mixing of the ocean and melted fresh waters.

This, in turn, may increase global warming and accelerate the approach of the next ice age.

Original source: Swansea University
Year: 2017

What do you think about scientific facts? It appears your opinion actually matters

QUICK BENEFITS: Think big, keep up with new information and use psychological tricks to convince science-skeptics

And not only yours - psychological researchers found out that people usually perceive facts based on their relevance to people's initial opinions.

Level of education doesn't matter. It's more about the lawyer-like thinking - when the person only focuses attention on the information that supports his beliefs.

Also, researchers discovered that the more scientific-curious is the person, the higher chances he'll be more open-minded towards scientific facts.

Psychologists recommend to find the core motivations of a person and then to frame scientific evidences based on them.

Original source: Society for Personality and Social Psychology
Year: 2017

NHTSA's NMVCC survey - 2-3 hours of sleep lead to doubled risk for a crash

QUICK BENEFITS: Sleep for 7-9 hours, don't drive when drowsy, use public transport if you had less than 6 hours of sleep

AAA Foundation' researchers analysed the data from thousands of police crash site reports. They found that if one sleeps 5-6 hours in a 24-hour period, he has a doubled chance for a crash while driving.

Here're the crash risk rates based on sleep hours:

6-7 hours:	1.3
5-6 hours:	1.9
4-5 hours:	**4.3**
3 and less:	**11.5**

Original source: AAA Foundation for Traffic Safety
Year: 2016

New Zealand's big geological problem - now explained

QUICK BENEFITS: If going to New Zealand, consider an earthquake possibility + avoid getting permanent residence there

An international team of geologists found hot water below New Zealand's Alpine Fault at 600m depth while boring a kilometre into it.

Usually, water at temperatures of more than 100°C is found in volcanically active regions like Iceland, Yellowstone USA, etc.

They also found that the ground below the Alpine Fault is moved up at the steady rate of 10mm per year, pushing hot rocks from 30km depth up.

Looking back at the geological history of New Zealand and these new findings, geologists expect a major earthquake - of 7-8 magnitude - in the next 50 years.

Original source: University of Southampton
Year: 2017

© John Townend, Victoria University, NZ

New optical wireless internet connection could replace Wi-Fi in 5 years

QUICK BENEFITS: The new era of safer & more efficient wireless technology might start soon - the optical one

Scientists developed a photonic data transfer method which reaches more than 40Gbit/s of download speed.

Infrared rays are emitted at safe frequencies and don't interfere with each other.

One infrared router could cover lots of connected devices so there shouldn't be any overload problems.

© wladimir1804 / Fotolia

© Vectorpocket / Freepik

But this technology has its own drawbacks - uploads can still be done by radio channels only, and the coverage of such router is relatively small.

Original source: Eindhoven University of Technology
Year: 2017

How do fears develop in children? It appears they can be adopted from parents and friends

QUICK BENEFITS: Teach children how to speak about their fears in a positive manner and to seek more encouragement

Researchers conducted a study on 242 British school children.

Children were shown pictures of unknown to them animals with both threatening and ambiguous descriptions. Then children were split into pairs so they could discuss the information they recently received.

Researchers found out that fear responses of children changed after the discussion: boy-boy pairs showed an increase in fear responses; however, girl-girl pairs showed a decrease.

These findings suggest that children should learn how to speak about their fears, especially with their anxious friends.

Original source: University of East Anglia
Year: 2016

CLAIM YOUR COUPON

for a 75% discount on the Udemy course

EMAIL DEVELOPMENT
FOR MARKETING AND ENTHUSIASTS

https://www.udemy.com/edemecourse/

EDEME4U-BESMART

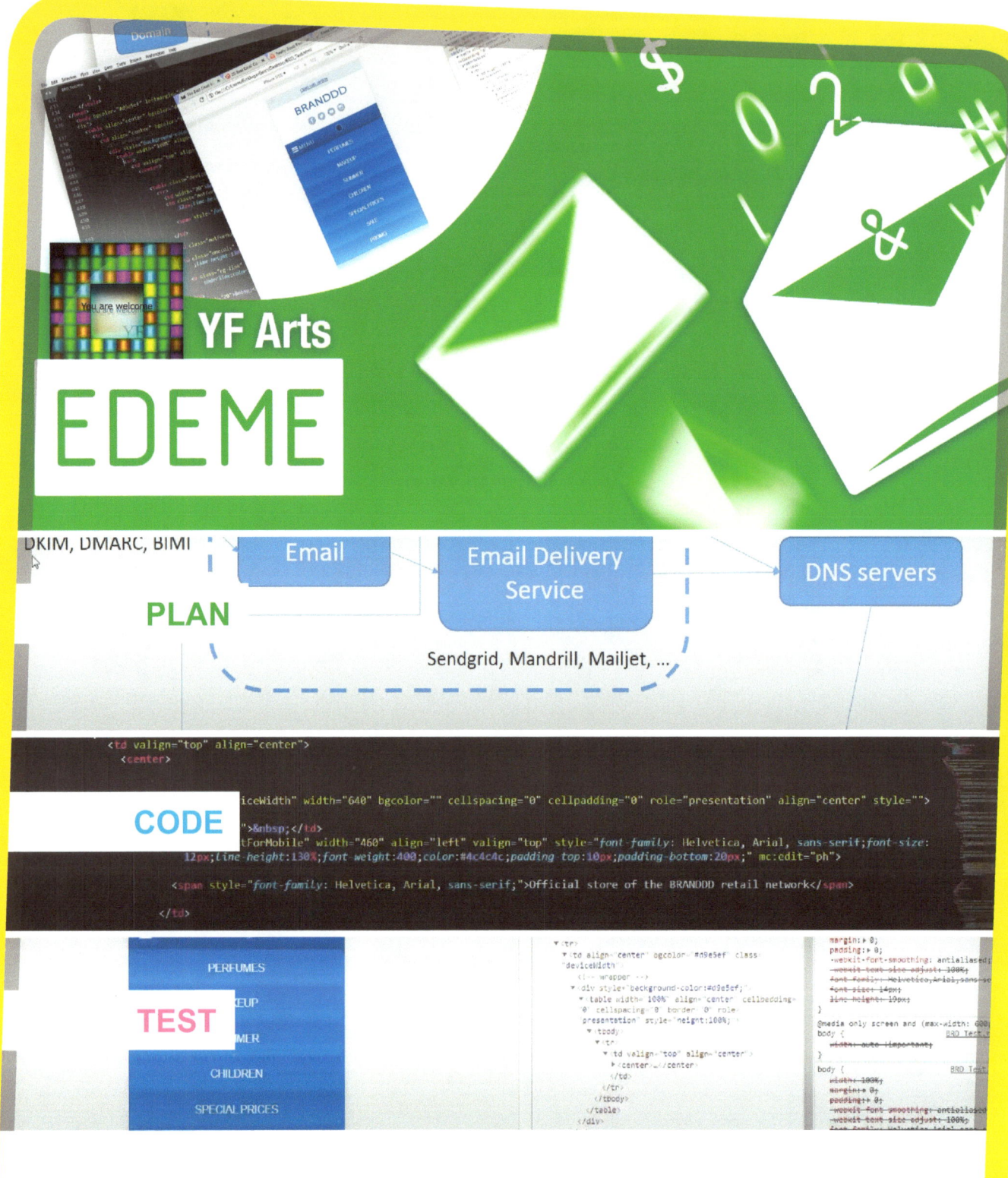

https://www.udemy.com/edemecourse/

Will you help others in life-threatening situations first? Or maybe yourself?

QUICK BENEFITS: In dangerous situations, save yourself first; after that, if proper equipment is present - try to help others

Actually, you wouldn't want to do that. The research on the computer model of a 3-level underground space showed that saving others first in a life-threatening situation led to higher casualties in the target group.

Different survival strategies were compared: without saving others, working with others as a team and saving others when already being in a safe place.

© Patsy Lynch / FEMA

© Welcomia / Freepik

As a result, the best survival strategy suggested by the research is to save yourself first and then try to help others from the safe place, only if you have proper equipment of course.

Original source: University of Waterloo
Year: 2017

13

Hit your 60th? No problem - you can still become a programmer

QUICK BENEFITS: Software development raises IQ in youth and postpones aging in the elderly. So why don't give it a try?

Research from the creator of the online teaching service shows that lots of online students are older people aged above 60.

Different motivations of those students were observed - keep the mind sharp, make up for missed opportunities, keep up with youth, develop professionally, implement a hobby idea, etc.

The findings suggest that programming should be more aligned with matters the student care about. Also, even in the ages over 60, one can still benefit from learning how to code.

Original source: University of California - San Diego
Year: 2017

Are you an impulsive person? If so, you might want to stay away from drugs

QUICK BENEFITS: Cognitive exercises and strengthening of working memory may help you or your children avoid drug abuse

Study shows that people with weak working memory are more likely to develop long-term drug abuse after initial experimentation at school age.

Some youth might start early and stop quickly, though others experiment and progress into heavier drug use.

© https://around.uoregon.edu

© Macrovector / Freepik

Scientists suggest that strengthening working memory and regular cognitive exercises in such adolescents may help them avoid drug usage and further abuse progression.

Original source: University of Oregon
Year: 2017

Even a few nuclear strikes will cause severe climate changes around the world

QUICK BENEFITS: Move to politically stable and water-rich regions of the world, and be prepared for long winters

Since the Cold War, lots of countries have started their own nuclear programs, and some of them, like North Korea, succeeded.

Scientists calculated the effects of nuclear missile strikes on climate models around the globe.

It appears that even 4-5 nuclear blasts may have a serious impact on the world we know.

The main risk factor is the delivery of millions of tons of ash and soot into the stratosphere during the nuclear blast.

It could cause global temperatures to drop to their 1,000-year minimum and up to 80% reduction in rainfalls in some areas of the world.

Original source: University of Nebraska-Lincoln
Year: 2017

Did you see those 4k or even 10k displays? Multiply these resolutions by a factor of 3

QUICK BENEFITS: Expect displays with 20-30k resolutions in stores in the next few years

Current LCD panels consist of millions of pixels, each of which is actually a group of 3 subpixels - Red, Green, and Blue -, in order to comply with the RGB color model.

Now researchers developed a new technique that allows RGB colors to be shown using only 1 subpixel, leading to up to 3x increase in display resolution.

© https://today.ucf.edu

© Starline / Freepik

The core principle lies in the modification of the nanostructure surface roughness under certain voltages. Nanostructure resembles an egg crate and is able to reflect light in full color range.

Original source: University of Central Florida
Year: 2017

How can Vitamin D be possibly associated with autism and pregnant women?

QUICK BENEFITS: For pregnant - walk 15+ minutes daily or/and Vitamin D supplements to prevent autism in newborns

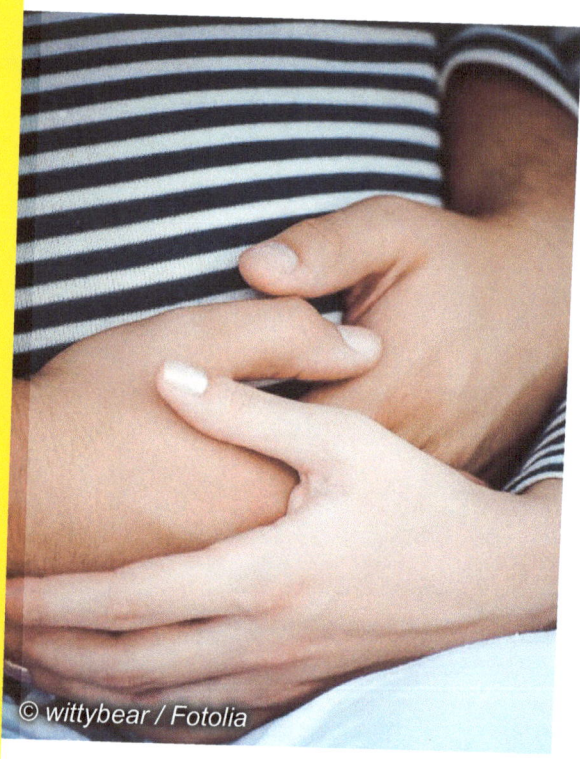

Study on mice revealed the connection between Vitamin D levels in pregnant women and brain development of the fetus.

Usually, Vitamin D is produced by skin cells while it's exposed to sun light.

Now researchers found out that low Vitamin D level in pregnant women is associated with high risks of the autism traits development in their children.

Mechanisms behind Vitamin D effect on the child's brain development are still mostly unknown.

Original source: University of Queensland
Year: 2017

Can we predict major tsunamis' occurrence, or can't we?

QUICK BENEFITS: When visiting coastal regions, monitor emergency warning channels and be prepared for evacuation

Since great 2004 Indian Ocean tsunami, scientists were constantly looking for places around the ocean coast to analyse the sediments for past tsunamis' signs.

Now, they found the oceanic L-shaped cave that contains sediments with signs of tsunamis between 7,900 and 2,900 years ago. More recent sediments were washed out by the 2004 great tsunami.

© Earth Observatory of Singapore

© Freepik

Based on their findings, tsunamis may occur with different frequency - once in 2 millennia, 4 times per millennia, etc. Scientists also found that major tsunamis follow smaller ones after long dormant periods.

Original source: Rutgers University
Year: 2016

Bots can affect news on the social networks - with both positive and negative impact

QUICK BENEFITS: Double-check news validity, control the information you post and use trusted online dating services

Researchers studied the way news become trending on Twitter. They found out that there're central accounts, which tweet or reshare the news catapulting them to "trending" status. And some of those accounts are actually bots.

Bots are able to "amplify" the message - they reshare it to lots of recipients, add special hashtags, and sometimes even able to chat, like the bots on some online dating websites...

Such bots could be used to raise awareness of some social events, protesting moves, employees' complaints, etc. But there's also a bad side - bots could easily be used to manipulate the mood of masses and push controversial reforms.

Original source: University of Georgia
Year: 2017

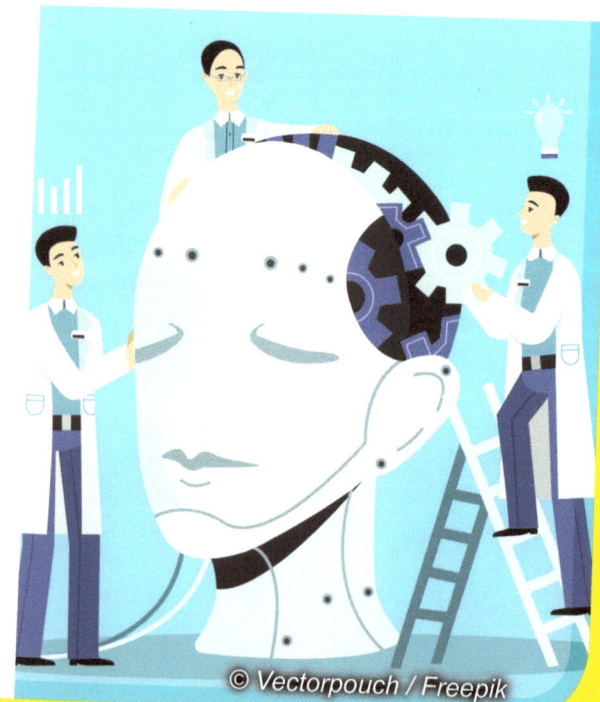

How does air temperature affect sleep quality?

QUICK BENEFITS: Try to keep your sleeping room temperature between 18 to 20 degrees Celsius to maintain good sleep quality

The large study's uncovered the relationship between air temperature and sleep quality in people of different age and income.

It appears that temperature rise by 1 degree Celsius over the regular one leads to 3 nights of insufficient sleep per 100 individuals per month. This translates into millions of lost sleep hours in the nationwide scale.

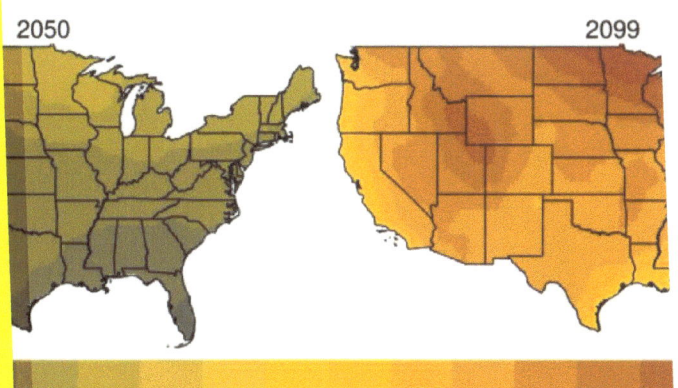

Based on these findings, researchers conclude that climate change may have more severe consequences than was thought before, including sleep deficiency among those with low-income or those aged 65 and older.

Original source: University of California - San Diego
Year: 2017

The Atlantic Meridional Overturning Circulation (AMOC) is at risk of disruption

QUICK BENEFITS: Move to live away from the Atlantic ocean coasts, or install a good climate control system

Did you know that AMOC, which includes the famous Gulf Stream, is critical for climate regulation in Atlantics? For the last few decades, AMOC strength has weakened by up to 15%, mostly because of the human-induced global warming.

Scientists studied different factors, which may lead to AMOC collapse. They found that ice melting is among the most critical of them as ice is melting really very fast in the Arctic and Greenland.

As AMOC plays a major role in heat distribution in Atlantics, such trends may lead to frequent freezing winters and scorching summers in Southern US and Western Europe & Africa.

Original source: Yale University
Year: 2017

Our brain unconsciously misjudges speed after it gets used to the current one

QUICK BENEFITS: Carefully exit high-speed motorways, rely on speedometers rather than on your own perception

You may have noticed that when driving with 70 mph speed your perception of the normal speed alters and you need time to adjust to slower movement.

Scientists examined the unconscious processes behind that phenomenon. They found out that our brain continuously compensates for changes in our visual stimulation. Thus we sometimes perceive higher speeds as normal.

This research could also be applied to sports judges, who may sometimes make non-objective decisions under such impact.

Original source: University of Lincoln
Year: 2017

Thunderstorms occur twice as frequenty over the areas with high aerosol emissions

QUICK BENEFITS: Expect higher rainfall and stronger storms when you travel via most popular ship routes

Researchers studied the major ship routes over some parts of the Indian Ocean and the South China Sea.

They discovered twice as much lightning over these routes compared to the ocean regions outside the ship routes.

Microscopic particulates - or aerosol -, emitted from ships, are actually the nuclei on which clouds form.

Small water droplets form on aerosol and then fly into the upper atmosphere, where they're condensed into the ice, which results into the lightning formation.

Original source: American Geophysical Union
Year: 2017

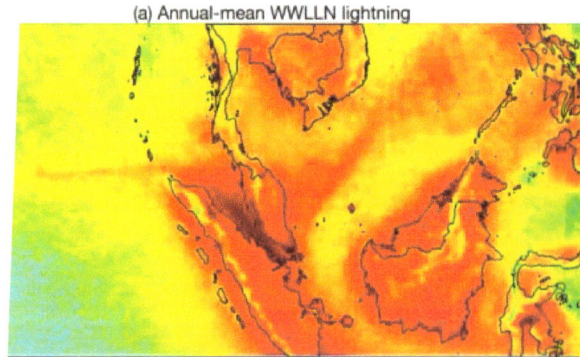

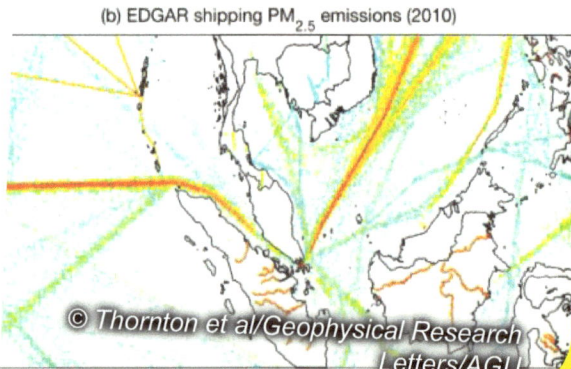

New AI that simulates social interaction in computer games is developed

QUICK BENEFITS: Prepare for computer games with complex social interactions - make sure your kids live the full social life

Researchers developed an AI, that adds non-linear social interactions for non-player characters (NPCs) in Skyrim - the well-known RPG game. The AI is available as a third-party modification on STEAM.

Now NPCs can execute different actions towards other NPCs or a player based on their knowledge of others' relationships between each other.

© Tamahikari Tammas

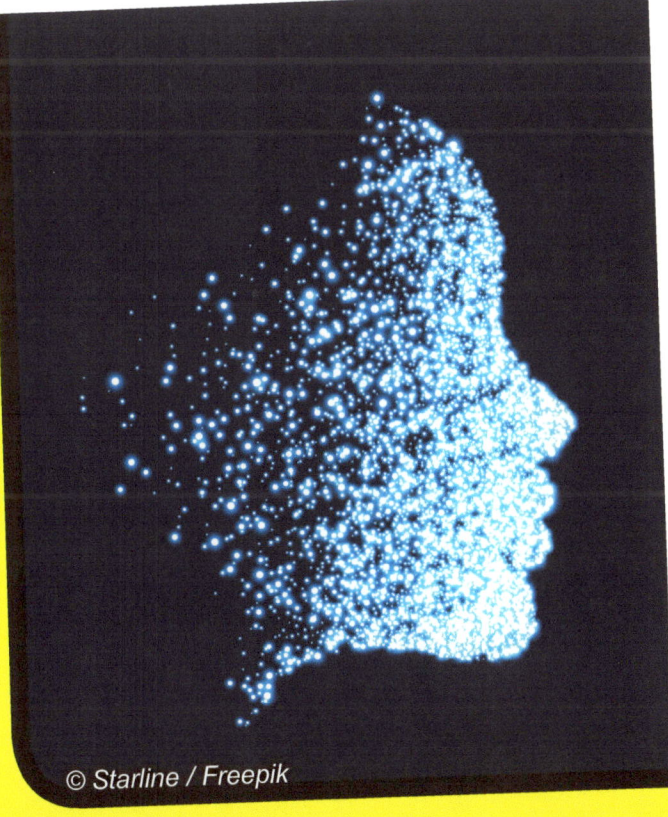
© Starline / Freepik

Currently, researchers are looking for partnership opportunities to built-in their AI into the next generation computer games.

Original source: North Carolina State University
Year: 2017

25

YF Arts
Design | UX | Emails

Instagram
yf_arts_

Dribbble, Quora
YF Arts

https://www.udemy.com/edemecourse/

All materials were taken from the official websites, dedicated to the corresponding studies.

All images were taken from public sources and are protected by corresponding copyrights. Additional copyrights for images from Freepik.com, used in the cover - © Fanjianhua, © Rawpixel.com, © Freepic.diller, © Freepik

Great thanks to the ScienceDaily.com for their work.